IT'S POSSIBLE

THE STORY OF RONALD MCNAIR

RITA LORRAINE HUBBARD

ILLUSTRATED BY CHASE WALKER

LITERARY SAFARI

Text by Rita Lorraine Hubbard
Illustrations by Chase Walker
© 2026 by National Medal of Honor Heritage Center

Academic Design by Maranda Wilkinson
Art Direction by Anjali Sakhrani
Design and **LP BLOCK** typeface by Carrie Lapolla
Edited by Sandhya Nankani
Fact Checking by Jahnavi Pradeep
Permissions by Kayla Fedeson
Photo Research by Scott Saunders

NASA imagery and insignia are used in this book for historical and educational purposes in accordance with NASA guidelines. Such use does not constitute official endorsement by NASA.

Illustrations include artist-created renderings referencing historical insignia of N.C. A&T and MIT, used with permission. All trademarks belong to their respective owners.

No AI-generated text or illustrations were used in the creation of this book.

The publication of this book was made possible by the National Medal of Honor Heritage Center which shares the stories of Medal of Honor and Space Medal of Honor Recipients with audiences of all ages. Support was provided by the Fields and Bell Families. www.mohhc.org

Library of Congress Control Number: 2025949696
ISBN: 979-8-9997231-4-7

Printed in the United States of America
First Impression 2026

Special thanks to Cheryl McNair and the estate of Ronald McNair.

THIS BOOK IS DEDICATED TO MY MOTHER,
WILLENA TIPSEY NUNN, WHO TAUGHT ME THAT NOTHING
IS IMPOSSIBLE IF YOU PUT YOUR MIND TO IT. — RLH

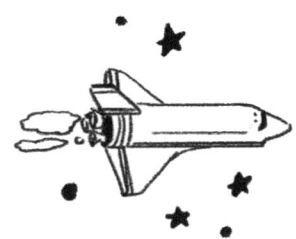

THIS IS DEDICATED TO AVA AND EDEN, WHO REMIND ME
EVERY DAY WHY I CREATE, AND TO MY WIFE, NERMIN, WHOSE
SUPPORT MAKES EVERYTHING POSSIBLE. — CW

Ronald McNair was a rock star.

He aced his classes.

LAKE CITY ELEMENTARY

Was a champ in sports.

Played the
saxophone
like a pro.

CARVER
HIGH
SCHOOL

CLASS
OF
1967

Became valedictorian.

He also earned a black belt in karate.

ki-Aii!

All of these were great accomplishments, but they weren't his greatest.

Ron made it to space. That's right, Ron actually went to outer space!
But before we join him among the stars, let's start at the beginning ...

Ron was born in a leaky little house in a rural little city in South Carolina. His father was a mechanic, and his mother was a teacher.

He and his brother Carl sweated in nearby tobacco and cotton fields to earn extra money. Some folks called his family "poor." Still, Ron never thought of himself as poor. He did not see poverty; he saw only possibilities. To Ron, life was an adventure, and he was going to enjoy every moment of it.

One of Ron's most important adventures happened the day he checked out two books from the library. This should not have been a big deal, but it was. You see, Ron lived during segregation, when Blacks were prevented from doing lots of things Whites could do. That included borrowing library books.

This awful rule didn't stop Ron. He chose the books he wanted, then stood firmly in line just like everyone else. When it was Ron's turn, the librarian scowled.

Her words stung, but Ron did not let go of the books.

The librarian called the police. Then she called Ron's mother. Ron still did not change his mind. Instead, he propped his nine-year-old body on the book counter. Slowly but surely, the whisper-quiet library turned into a storm … all over a couple of library books!

Ron never gave in. But the librarian finally did. This might have been the day Ron learned he could do anything he put his mind to.

Ron was just like everyone else, but he was different too.

He loved unusual things, like Sputnik, the Earth's first official satellite.

He also loved *Star Trek*, the popular new television series. His favorite character was Lieutenant Uhura, the communications officer on the show. Uhura was smart; she was Black; and she was right where Ron wanted to be ... among the stars.

What's it like up there? he wondered.

Going into space was a different kind of dream. No one Ron knew had ever dared to dream so high. Even his brother Carl did not understand.

SO, HOW IS A COLORED BOY FROM SOUTH CAROLINA WHO WEARS GLASSES AND NEVER FLEW A PLANE ... HOW IS HE GONNA BECOME AN ASTRONAUT?

It seemed impossible. But it wasn't. Ron only had to do three things to make it happen: He had to study hard. He had to believe in himself. And he had to set his sights on the stars.

So … Ron studied.

Sometimes he doubted himself.

When that happened, he welcomed encouragement.

He celebrated ...
and he studied some more.

And before he knew it, he was graduating with a PhD.

MIT

It was time. Ron was ready to shoot for the moon ... literally!

Could he really become an astronaut? Did he have what it took?

He wasn't sure. But he had worked hard and studied even harder. He wasn't going to back out now.

He sent in his application.

And what do you know?
Ron made it!

CONGRATULATIONS, RONALD MCNAIR.
YOU'RE GOING TO BE PART OF THE NEXT
GENERATION OF ASTRONAUTS!

NASA SELECTS
35 OF 8000
APPLICANTS

Over the next few years, Ron trained and prepared.

And then, on February 3, 1984, it happened! A **RUMBLE** … and a **ROAR**. A mushroom of smoke and flames.

Vibrations, and then weightlessness …

Ron had made it.

He was in space.

Ron's life had not always been easy. But through it all, he had shown integrity, citizenship, patriotism, and sacrifice. He had let commitment and courage guide his every step.

The boy who began life with very little never let go of his belief that he would make it to space one day.

Welcome, Ron.

The stars have been waiting for you.

A NOTE FROM THE AUTHOR

October 21, 1950 - January 28, 1986

RONALD ERWIN MCNAIR was born on October 21, 1950 in Lake City, South Carolina. He studied Physics at North Carolina A&T University before receiving his PhD in Physics from MIT.

Ron's first flight, from February 3 to February 11, 1984, was aboard the Space Shuttle Challenger. He told his wife, Mrs. Cheryl McNair, that the sight of the Earth in the darkness was like seeing an oasis in the desert.

During this flight, Ron performed experiments, analyzed cancer cells, and studied how zero gravity affected arthritis. He also served as the flight's official photographer and cinematographer. Ron took his saxophone with him to play a medley of songs. According to Mrs. McNair, the first song, *America the Beautiful*, was directed toward every man, woman, and child on the face of the Earth. The second song, *What the World Needs Now is Love Sweet Love*, focused on how humans have the power to correct some of the things going awry in the world through love and care. And the third song, *Reach Out and Touch Somebody's Hand*, focused on how we can make the world a better place.

Ron's second flight was scheduled to occur on the morning of January 28, 1986, aboard the same Space Shuttle Challenger. Tragically, the Challenger exploded 73 seconds into the flight. Commander Francis R. (Dick) Scobee, Pilot Michael Smith, Payload Specialists Sharon Christa McAuliffe & Gregory Jarvis, and Mission Specialists Judith A. Resnik, Ellison S. Onizuka, & Ronald McNair were all lost in the explosion.

For this second flight, Ron had planned to be the first person to record music in space. He collaborated with French composer Jean-Michel Jarre to compose a saxophone solo. The plan was to broadcast the performance live from orbit, to accompany Jarre's *Rendezvous Houston* concert on Earth. The performance never took place.

To honor Ron, Jarre named the song they were supposed to perform together *Last Rendezvous (Ron's Piece),* and French jazz musician Pierre Gossez played Ron's part. Jarre said, "Ron was so excited about the piece that he rehearsed it continuously until the last moment. May the memory of my friend, the astronaut and the artist, Ron McNair live on through this piece."

On July 23, 2004, Ronald McNair and the crew of the Space Shuttle Challenger received the Congressional Space Medal of Honor. This medal has only been awarded a few times and was designed to honor "any astronaut who in the performance of his duties has distinguished himself by exceptionally meritorious efforts and contributions to the welfare of the Nation and of mankind." The award was presented by former President George W. Bush.

Today, many schools and organizations throughout the United States are named after Ronald McNair. Carver High School—the school Ron attended—is now called Dr. Ronald E. McNair Junior High School.

ABOUT THE NATIONAL MEDAL OF HONOR HERITAGE CENTER

Located in Chattanooga, Tennessee—the "Birthplace of the Medal of Honor"—the **National Medal of Honor Heritage Center** shares the BIG stories of Medal of Honor and Space Medal of Honor Recipients with audiences of all ages. These awards are the highest military and space decorations the United States of America bestows. Ronald McNair was awarded the Space Medal of Honor in 2004.

The stories of Medal of Honor and Space Medal of Honor Recipients capture the spirit of what it means to put service before self and to embrace curiosity. They remind us all individuals, while ordinary, are capable of the extraordinary.

The values of the Heritage Center are centered around the six character traits embodied by the Medal of Honor and all of its Recipients: Patriotism, Citizenship, Courage, Integrity, Sacrifice, and Commitment. Through the Heritage Center's immersive onsite galleries and BIG energy educational initiatives and experiences, students, teachers, and the general public are connected to this rich history.

www.mohhc.org | Instagram @thefirstmedals | Facebook @thefirstmedals

This literary work was made possible through the generous support of the Fields and Bell Families.

ABOUT THE AUTHOR

RITA LORRAINE HUBBARD is a former teacher of 20+ years. She is the author of *The Oldest Student: How Mary Walker Learned to Read, Hammering for Freedom: The William Lewis Story,* and *African Americans of Chattanooga: A History of Unsung Heroes.* Her books have won multiple awards including the Texas Bluebonnet Award. She lives in Chattanooga, Tennessee, with her dogs Rocky and Joey.

www.ritahubbard.com

ABOUT THE ILLUSTRATOR

CHASE WALKER is a self-taught illustrator and artist. Born in Liberia, West Africa, his early life was marked by the turmoil of civil war. Surrounded by hardship at a refugee camp in Ghana, he found hope by making art in the sand. Today, he lives in Central Pennsylvania with his wife and two young daughters and continues to tell stories through his art that bridge his past and present. His artwork is part of the National Museum of American History collection.

www.iamchase.co